UNDERSTANDING

WILDERNESS

THERAPY

I0454387

Unlocking Nature's Healing Power: A Deep
Dive Into The Principles And Benefits For
Holistic Well-Being, Personal Growth, Mental
Wellness And More

DR. KARSON BRYAN

Copyright © 2023, Dr. Karson Bryan

Reserved rights apply. Except for brief quotations included in critical reviews and certain other noncommercial uses allowed by copyright law, no part of this publication may be reproduced, distributed, or transmitted in any form or by any means, including photocopying, recording, or other electronic or mechanical methods, without the publisher's prior written permission.. In order to enable others to ask for permission for particular purposes, you can also include your contact details. To make sure that your copyright statement conforms with all applicable rules and regulations, you might also choose to speak with a legal expert or copyright specialist.

DISCLAIMER

This book's content is meant to be used solely for general informative purposes. Despite having taken every precaution to guarantee the content's accuracy, the author disclaims all duty and responsibility for any errors or omissions. It is recommended that readers exercise caution and, if needed, seek expert guidance. Any and all liability for losses, damages, or other outcomes arising from the use of the material included in this book is disclaimed by the author and publisher. All referenced product names and trademarks are the property of their respective owners and are merely cited for identification. Any likeness to real people or things is entirely accidental. Since it is a work of fiction, this book should not be used as a substitute for professional, legal, or medical advice. It is advised that readers seek advice on particular issues from qualified experts."

Please make sure that this disclaimer is modified to fit the particular requirements and subject matter of your book. Seeking advice from a legal expert is also a smart option if you have any questions or require a more thorough disclaimer for your specific book.

TABLE OF CONTENTS

CHAPTER ONE

WILDERNESS THERAPY

INTRODUCTION

In the outdoor setting, wilderness therapy offers a dynamic and all-encompassing approach to healing, personal development, and self-discovery. The unique fusion of outdoor adventure, experiential learning, and mental health treatment has made this therapeutic method more and more well-known in recent years. This thorough introduction seeks to explore the fundamental ideas of wilderness therapy, offering a thorough grasp of its definition, historical evolution, and the many advantages it provides to anyone looking for a life-changing experience in the great outdoors.

WILDERNESS THERAPY: WHAT IS IT?

To promote personal development and emotional recovery, those who receive wilderness therapy—also known as adventure therapy or outdoor behavioral healthcare—are placed in a wilderness environment. It is predicated on the idea that positive change, emotional processing, and self-discovery may all be greatly accelerated by exposure to the natural world. In wilderness treatment programs, participants usually live and work for a considerable amount of time in isolated, natural environments, participating in a range of outdoor activities such as hiking, camping, rock climbing, and backpacking. Trained therapists and outdoor specialists lead these excursions, assisting participants in navigating the opportunities and difficulties that the outdoor environment presents.

THE DEVELOPMENT AND HISTORY OF WILDERNESS THERAPY

Back in the middle of the 20th century, outdoor programs were initially employed as an alternate method of helping problematic kids, which is when wilderness therapy originally emerged. The idea of using the environment as a therapeutic setting was established in the United States by pioneers like Gene Wilkes and Larry Dean Olsen. In 1967, Olsen wrote a book titled "Outdoor Survival Skills," which emphasized how the outdoors can impart important life lessons and boost self-esteem.

The field of wilderness therapy has developed and broadened throughout time to serve a wider spectrum of clients, including adults, those overcoming addiction, and those pursuing personal development and self-discovery. The integration of evidence-based therapy techniques and the formation of industry standards and accreditation organizations have resulted in a more structured and professional approach. Programs for wilderness treatment are now offered in a variety of formats, ranging from brief interventions to extensive residential settings, all tailored to meet the individual needs and therapeutic objectives of the client.

ADVANTAGES OF USING WILDERNESS THERAPY

With so many advantages, wilderness therapy is an appealing option for anyone looking to grow personally and heal emotionally. The natural world is first and foremost therapeutic, offering a calm and breathtaking backdrop that promotes introspection and a sense of connection with one's surroundings. Participants are free to concentrate on themselves and their therapeutic journey because there are no contemporary distractions present, such as electronic devices or the never-ending demands of daily life.

Furthermore, the physical demands of outdoor sports can foster resilience, self-worth, and a feeling of achievement. Overcoming unavoidable challenges and taking part in group activities promotes a sense of community and cooperation, which enhances social skills and interpersonal connections. Through the process of self-discovery, the therapeutic alliance with qualified experts supports individuals as they face and work through emotional problems, traumatic experiences, and personal challenges.

Furthermore, the qualities learned in the environment can be directly applied to daily life, improving one's capacity for problem-solving and judgment. In addition to encouraging a strong bond with the natural world, wilderness therapy promotes environmental consciousness and respect for it.

Wilderness therapy is a revolutionary method for promoting personal development and healing that blends the therapeutic knowledge of qualified specialists with the potent experiences provided by the natural world. Anyone thinking about this unusual and successful kind of treatment must comprehend the idea of wilderness therapy, its historical origins, and the many advantages it offers. Wilderness therapy provides a route to transformation in the great outdoors, whether one is looking to overcome personal obstacles, develop resilience, or go on a self-discovery adventure.

CHAPTER TWO

THE BASICS OF THERAPY IN THE WILDERNESS

NATURE AND HEALING

People have long acknowledged the therapeutic benefits of nature, which provide them with a deep sense of comfort and renewal. The intrinsic link between healing and nature serves as the cornerstone of wilderness therapy. With its tranquil vistas and pristine beauty, the natural world possesses a special power to calm the mind, lessen tension, and enhance mental well-being. It gives people a place to escape the pressures of contemporary life and re-establish a connection with their environment and selves.

THE HEALING POWER OF NATURE

An increasing amount of scientific study backs up the anecdotal evidence of nature's healing abilities. Time spent in natural settings has been shown to improve mood, lower cortisol levels, and lessen anxiety. Natural environments, such as peaceful forests, flowing rivers, or breathtaking mountain ranges, have a relaxing effect that can assist people in escaping the everyday stresses that frequently result in mental health issues. People can discover balance and serenity in nature, which provides a break from the hectic pace of modern life.

ECOPSYCHOLOGY AND WILDERNESS THERAPY

The study of the interaction between human psychology and the natural environment is known as ecopsychology. It recognizes the detrimental effects on mental health of modern society's alienation from nature. Based on ecopsychology, wilderness therapy addresses these problems by using nature as a therapeutic medium. Ecopsychology-based wilderness treatment assists people in re-establishing their connection to nature by placing them in a natural setting.

This reconciliation can promote emotional healing, self-awareness, and personal development.

RELATIONSHIP BETWEEN NATURE AND MENTAL HEALTH

There are several facets to the relationship between nature and mental health. Stress, anxiety, and depressive symptoms can all be lessened by spending time in natural settings. Being in nature encourages introspection and awareness, which calms the mind. Furthermore, the physical pursuits that are frequently linked to wilderness treatment, such as hiking, camping, and outdoor adventure, can increase confidence and self-esteem while offering a much-needed sense of success.

The goal of wilderness therapy is to assist clients in overcoming their emotional and psychological obstacles by acknowledging the inextricable connection between mental health and the natural world. It invites customers to interact with nature and take inspiration from its tenacity, rhythms, and adaptability. In doing so, people might enhance their mental health by developing coping mechanisms, a greater sense of self-awareness, and a deeper appreciation for their surroundings.

Wilderness therapy promotes mental health and personal development by relying on the profound relationship between healing and the environment. The therapeutic and scientifically validated healing properties of nature make it an invaluable tool for tackling the problems of contemporary living. This method is supported by ecopsychology, which emphasizes the value of reestablishing a connection with nature to enhance mental health. Through the recognition and utilization of the relationship between mental health and nature, wilderness therapy provides a comprehensive and efficacious method for healing and personal development.

CHAPTER THREE

THERAPEUTIC METHODS

MODELS OF WILDERNESS THERAPY

Therapeutic interventions can be approached in a novel and immersive way with the help of wilderness therapy models. These methods are based on the idea that people with a variety of emotional and behavioral difficulties might find a healing and transformative setting in nature. Participants in wilderness treatment are removed from their typical comfort zones and placed in natural environments, such as deserts, mountains, or woods. The natural difficulties and unpredictability of the outdoors present chances for introspection, personal development, and therapeutic breakthroughs. To foster self-discovery and resilience, wilderness treatment models frequently include outdoor activities such as hiking, camping, and survival skills training. These models highlight the significance of the natural environment in promoting personal and emotional growth.

ADVENTURE-BASED COUNSELLING

A closely related method to wilderness therapy is called adventure-based therapy, which emphasizes outdoor experiences and adventure as catalysts for therapeutic transformation. Adventure-based therapy programs enable participants to face phobias, increase self-esteem, and learn problem-solving techniques through engaging in activities like rock climbing, kayaking, and ropes courses. These experiences are meant to push people beyond their comfort zones while providing a supportive and supervised setting. Through the development of coping mechanisms, teamwork, and communication, adventure-based therapy promotes personal growth. Participants frequently report feeling more empowered and confident, which can have a long-lasting effect on their lives.

METHODS THAT ARE COGNITIVE-BEHAVIORAL

A popular type of therapy that focuses on the interaction between ideas, feelings, and behaviors is called cognitive-behavioral approaches. Using this method, therapists assist clients in recognizing and changing harmful thought patterns and beliefs that underlie emotional suffering or troublesome behaviors. Cognitive-behavioral therapy (CBT) is a highly structured, goal-oriented treatment approach that teaches people useful coping mechanisms and problem-solving techniques. CBT is beneficial for treating a variety of mental health conditions, such as addiction, depression, and anxiety. It helps people identify their cognitive distortions and replace them with more adaptive and logical ideas, which improve behavior and emotional health.

GROUP PROCESSES AND COUNSELING

Group dynamics and therapy are essential components of many therapeutic modalities. In group therapy, a therapist leads a small group of people dealing with related problems during a session. The group's interactions and dynamics can be therapeutic in and of themselves, giving members the chance to talk about their experiences, get support, and get input from other members. In addition to lowering feelings of loneliness and fostering social skills and interpersonal development, group therapy can provide a feeling of acceptance and affirmation. Various group therapy modalities are designed to address certain needs and therapeutic goals, including process-oriented groups, psychoeducational groups, and support groups. Because people may practice new skills in a social setting and learn from each other, group dynamics can improve the therapeutic process.

CHAPTER FOUR

MORAL DETERMINATIONS

PRINCIPLES OF ETHICS IN WILDERNESS THERAPY

A novel and becoming more and more well-liked method of assisting people with a range of emotional and behavioral problems is wilderness therapy. To protect participants' safety and well-being, it must, nevertheless, abide by ethical guidelines just like any other type of therapy. The use of wilderness treatment is guided by several important ethical precepts.

Primarily, the principle of beneficence—which prioritizes the participants' welfare and best interests—is essential to wilderness therapy. In addition to making sure that participants' growth and development are aided by the therapeutic process, therapists and staff must put participants' physical and emotional safety first.

This entails preventing injury while offering the proper resources, support, and interventions.

Autonomy is a crucial ethical factor in wilderness therapy. It is imperative to provide participants with the means to make well-informed decisions regarding their involvement in the program. This principle is closely linked to the idea of informed consent.

EDUCATED CONSENT

A fundamental ethical tenet in both therapy and healthcare, informed consent plays an equally important role in wilderness treatment. It denotes that participants have the right to complete disclosure regarding the nature of the therapeutic program, its possible advantages and disadvantages, and the particular techniques used. With informed consent, people are protected from being coerced or manipulated into making decisions about their engagement.

Obtaining informed consent in the context of wilderness treatment entails giving comprehensive and easily understood information about the program, including its goals, procedures, and possible hazards. The chance to ask questions and receive satisfactory answers must be provided to participants and, if applicable, their legal guardians. It should be possible to revoke consent at any moment and without consequence.

Because wilderness treatment participants frequently encounter difficult and potentially dangerous outdoor conditions, informed consent is especially important in this context. They must be aware of the potential emotional and physical difficulties and give their agreement by their comprehension of the program.

RISK AND SAFETY MANAGEMENT

One of the most important ethical rules in wilderness therapy is participant safety. Participants run the chance of experiencing a variety of environmental hazards, such as interactions with wildlife and bad weather, while severe weather may be present. Thus, risk management is essential to upholding moral principles in this industry.

It is the responsibility of therapists and program personnel to identify and reduce any dangers that may be present in the wilderness setting. This entails offering the proper tools, instruction, and oversight to reduce the likelihood of mishaps or injuries. Plans for emergency reactions must exist, and participants must get training on how to handle possible threats.

Finding a balance between the therapeutic advantages of wilderness therapy and the participants' safety is crucial. Programs that uphold ethics ought to push participants as far as they can go while making sure that the risks are appropriately handled.

CHAPTER FIVE

THE ENCOUNTER IN THE WILDERNESS

SELECTING THE IDEAL WILDERNESS ENVIRONMENT

Whether you're organizing a long-term backcountry expedition or a weekend camping trip, picking the right wilderness setting is essential. The journey's success, safety, and overall experience can all be significantly impacted by the destination selection. Several things need to be carefully evaluated to make an informed conclusion.

CHOOSING A LOCATION

The procedure starts with location selection, which is the most important phase. The kind of wilderness location that best suits an individual's goals and tastes must be chosen. Which kind of experience would you rather have—a quiet, isolated wilderness retreat or a well-known, crowded area with all the conveniences needed? It's critical to specify your travel objectives and the type of experience you hope to have. Your decision should be influenced by elements like recreational possibilities, natural characteristics, and distance from home.

THE ENVIRONMENT IN MIND

When choosing a wilderness responsibly, environmental considerations are extremely important. It is imperative to investigate the ecological relevance of the selected location to uphold and preserve the natural ecosystem. Certain wilderness areas may have delicate ecosystems, so you should take particular care and safeguards to lessen your influence. It is essential to comprehend local weather patterns, flora and fauna, and endangered species when organizing an environmentally conscious journey.

LOGISTICS AND ACCESS

Practical factors like logistics and access can make or ruin a trip into the bush. Access is how you will travel to your destination, be it via automobile, boat, foot, or even air. How easily you can get to the location will determine how many other outdoor enthusiasts you run with and how much privacy you want. You should also think about the trip's logistics, including transportation, equipment, and permissions. Certain wilderness regions could have stringent rules or permit requirements, while others might not have basic amenities like emergency services or sources of drinkable water.

Thorough research on all these aspects is vital before setting out on a wilderness excursion. Examine official park websites, peruse guidebooks, speak with knowledgeable outdoor lovers, and consider local experts' advice. In the end, the ideal wilderness setting should complement your objectives and standards, honor the natural world, and take into account the location's logistical difficulties. You can increase your chances of having a safe, fun, and rewarding wilderness adventure by carefully weighing location selection, environmental considerations, access, and logistics.

CHAPTER SIX

TOOLS AND DEVICES

ESSENTIAL EQUIPMENT FOR REWILDING

Encouraging and transforming, wilderness therapy is based on the healing advantages of nature and outdoor activities for mental and emotional rehabilitation. Being well-prepared and equipped with the appropriate equipment is essential for a successful and safe wilderness treatment session. In this article, we'll look at the necessary equipment for wilderness treatment and talk about how important it is to plan and pack, as well as how important safety equipment is.

ORGANIZING AND GETTING READY

The core components of a good outdoor therapy trip are careful planning and packing. An easy travel can be distinguished from a difficult one by careful preparation and arrangement. Before anything else, participants and facilitators need to evaluate the particular needs of their trip, considering elements like destination, length, and time of year.

Important things to think about are:

1. Clothes: It's important to wear layers to adapt to the shifting weather. This needs to have insulating mid-layers, waterproof outer layers, and moisture-wicking base layers. Remember to wear appropriate shoes, like well-made hiking boots.

2. Shelter: For a restful night's sleep in the woods, you'll need tents, sleeping bags, and sleeping pads. The season and climate of the therapy area should be taken into consideration while selecting a shelter.

3. Food and Water: Enough food and water supplies, as well as a purification technique, need to be packed. It is crucial to plan and prepare meals that are portable and simple to prepare when out in the field.

4. Navigational Aids: For safe navigation in the backcountry, maps, compasses, GPS units, and the ability to utilize them are essential.

5. Backpacks: Pick a backpack that fits properly and can accommodate all the equipment you'll need. Make sure the weight distribution is adjusted accordingly.

6. First Aid pack: Mishaps may occur, and to treat minor wounds and handle medical emergencies on the job site, a thorough first aid pack is essential.

7. Personal Items: It's important to remember to pack personal hygiene products, prescription drugs, and any materials connected to a particular therapy.

SAFETY SUPPLIES

Since wilderness treatment frequently entails traveling into isolated and possibly dangerous areas, safety gear is an essential part of the experience. The purpose of safety equipment is to reduce hazards and guarantee the health and safety of facilitators and participants.

Among the crucial safety gear are:

1. Communication Devices: Satellite phones or two-way radios may be necessary for group coordination and emergency communication, depending on the area and cellular availability.

2. Lighting: To navigate in poor light and to feel safe at night, headlamps and flashlights are essential.

3. Signaling Devices: In an emergency, whistles, signal mirrors, and flares are useful instruments for warning other group members or rescuers.

4. Emergency Shelter: In case of unforeseen bad weather, small, light emergency shelters such as space blankets or bivvy bags can offer protection.

5. Tools for Making Fires: Tools for starting fires, including waterproof matches or a fire starter, are

essential for cooking, staying warm, and indicating assistance.

6. Protective Gear: Depending on the situation, safety during particular activities may require the use of protective gear such as helmets, gloves, and eyewear.

7. Safety Guidelines: Just as important as the equipment itself is training in emergency procedures and wilderness safety measures. Both facilitators and participants need to be familiar with the procedures to be followed in different situations.

Wilderness therapy is an amazing therapeutic method, but it necessitates careful planning and the usage of the necessary equipment to guarantee everyone's comfort and safety. Having the appropriate gear, packing carefully, and paying attention to safety precautions are essential for a productive and life-changing wilderness therapy session.

CHAPTER SEVEN

SENSE OF NATURE AND SURVIVAL EXPERTISE

EDUCATING ABOUT NATURE AWARENESS

Building a stronger bond between people and the natural environment requires teaching nature awareness. It entails learning about the surroundings, its cycles, and the complex interactions that keep life on Earth alive. This educational process encompasses more than just imparting knowledge; it also involves fostering an appreciation, curiosity, and sense of duty toward the natural world. People are commonly taught nature awareness through experiential learning, which encourages them to explore and interact with the outdoors. It includes a variety of skills, including tracking, identification of plants, bird language, and ecological understanding.

The concept of the "sit spot," which is a particular place in nature that a person frequently visits to observe and connect with the environment, is a fundamental component of teaching environmental awareness. Engaging in this exercise enhances one's comprehension of the dynamic natural environment and fosters a feeling of place. It also strengthens one's sense of connectedness to the environment by encouraging attention and the capacity to detect little changes.

ESSENTIAL SURVIVAL KNOWLEDGE

The essential skills needed to survive and travel in the wilderness are known as basic survival skills. These abilities cover a broad spectrum of information and methods, such as creating a shelter and lighting a fire, locating and sanitizing water, recognizing edible plants, and performing basic first aid. Anyone who spends time outside has to learn and become proficient in these abilities since they can be the difference between life and death in a survival scenario.

DEVELOPING FORTITUDE IN THE BACKCOUNTRY

The goal of "Building Resilience in the Wilderness" is to acquire the mental, physical, and emotional toughness necessary to handle obstacles and crises encountered when hiking in the outdoors. In addition to learning how to deal with discomfort, adjust to the unpredictable nature of the wild, and have an optimistic outlook, developing survival skills is one way to develop resilience. Furthermore, to be resilient in the wilderness, people must become highly adaptive, cultivate self-awareness, and recognize their limitations.

Building resilience requires not just practical abilities but also an understanding of the psychological and emotional components of life. People need to learn how to control their emotions since in a survival crisis, tension, panic, and fear can be harmful. This can be accomplished by instruction, practice, and exposure to controlled stressors, which will equip people to handle difficult situations with poise and sound judgment.

Developing wilderness resilience, learning basic survival skills, and teaching nature awareness are all related activities that improve our connection to and capacity for living in the natural world. These ideas stress the significance of creating a strong bond with the natural world, learning vital survival skills, and equipping people to deal with the difficulties and uncertainties of the wild. When combined, they offer a comprehensive method for outdoor learning and living in the wilderness that guarantees not just survival but also a deep understanding of the wonder and intricacy of the natural world.

CHAPTER EIGHT

PARTICIPANTS AS WELL AS GROUPS

YOUTH & ADOLESCENT THERAPY

This particular type of mental health care focuses on treating the emotional, psychological, and behavioral issues that young people—usually those in early childhood to late adolescence—face. This type of treatment adapts interventions to fit the particular demands of the age group by acknowledging the developmental challenges and needs that are distinct to them. Adolescents frequently struggle with problems including identity formation, peer pressure, academic stress, and family conflicts, so providing therapy help that is tailored to their needs is essential.

The purpose of youth and adolescent therapy is to offer a secure and encouraging setting where young people can communicate their ideas and emotions, learn coping mechanisms, and grow resilient and self-aware. Depending on the needs of the person, therapists working with this demographic may use a range of therapeutic modalities, such as play therapy, family therapy, cognitive-behavioral therapy, and more. Better coping mechanisms, the capacity to forge stronger relationships, and gains in emotional well-being are frequently used to gauge how successful this type of therapy is.

WILDERNESS THERAPY FOR PROBLEMATIC YOUTH

Treating problematic youth with wilderness therapy is a novel and unorthodox method. It entails removing teenagers from the demands and distractions of daily life and placing them in a natural setting, where they can use the difficulties of wilderness experiences to promote personal development and healing. Through activities like camping, hiking, and survival skills training, this treatment seeks to instill self-reliance, self-esteem, and a sense of success in its participants.

The foundation of the wilderness therapy approach is the idea that the natural environment offers a special environment for introspection and personal growth. During their stay in the woods, participants frequently face their fears, learn how to solve problems, and form close relationships with peers and therapists. This type of therapy can be very helpful for difficult young people who are struggling with emotional disorders, behavioral challenges, and substance usage.

PROGRAM DESIGN AND STRATEGY

The effectiveness of youth and adolescent treatment programs is greatly influenced by the design and strategy of these programs. These programs need to be customized while keeping in mind individual characteristics to meet the unique demands of the target group. A well-designed program usually starts with a thorough assessment to pinpoint each participant's unique challenges and assets. Based on this assessment, the best therapy methods are taken into consideration while developing the treatment plan.

Programs for youth treatment may employ a variety of approaches, such as family therapy, group therapy, and individual counseling. The secret is to establish a safe, nurturing space where people may examine their feelings and develop good coping mechanisms. To offer a complete support system, therapists frequently collaborate with other specialists like social workers, teachers, and school counselors.

CASE STUDIES

Research and clinical practice in the field of youth and adolescent treatment depend heavily on case studies. Therapists and researchers can obtain useful insights into effective tactics and best practices by reading through the extensive reports of individual experiences, therapeutic interventions, and treatment outcomes that they provide. Case studies frequently highlight the complexity and diversity of the problems that young people confront as well as the various ways that therapy can improve their lives.

Therapists can make educated decisions regarding treatment plans for their clients, adjust their methods, and gain insight from real-life instances by looking at case studies. Furthermore, case studies can be used by researchers to produce hypotheses and provide guidance for the creation of evidence-based procedures. They present a complex picture of the difficulties encountered by children and teenagers, highlighting the possibility of development and improvement with therapy.

CHAPTER NINE

GROUP AND ADULT COUNSELING

ADULT WILDERNESS THERAPY

Adults who seek mental health and personal development may find that wilderness therapy is an exceptionally effective and distinctive method. It entails bringing people into outside, natural environments so they can participate in therapeutic activities and establish a connection with the surroundings. This kind of treatment is very well-liked by adults who are dealing with a range of problems, such as substance misuse, stress, anxiety, and despair. The opportunity to escape the rigors and distractions of contemporary life in the forest gives people the ability to reframe their problems and experiment with different coping mechanisms.

Adult participants in wilderness treatment programs frequently engage in outdoor pursuits like hiking, camping, and group talks. These exercises promote introspection and personal development, which strengthens resilience and fosters independence in people. Trained professionals facilitate the therapy process by assisting participants in overcoming emotional and psychological obstacles. Participants can gain transferable skills including communication, problem-solving, and emotional control.

TREATMENT FOR SUBSTANCE ABUSE AND ADDICTION

Given that addiction is a widespread and dangerous problem that affects millions of adults worldwide, substance abuse and addiction treatment is an essential part of adult therapy. Treatment strategies that are effective for addiction and substance abuse frequently combine counseling, therapy, and support. Treatment sessions, both individual and group, are important for treating the many psychological and social aspects that lead to addiction.

Adults who are abusing substances can address the root causes of their addiction, create coping mechanisms, and establish recovery objectives through individual therapy. It offers a private, secure setting for talking about personal matters and getting an understanding of behavioral patterns. On the other hand, group therapy provides a beneficial support system where people may talk about their experiences, encourage one another, and get knowledge from others going through similar struggles. Through the creation of a sense of community and a reduction in feelings of isolation, group therapy can promote a common commitment to recovery.

GROUP COHESION AND ASSISTANCE

Essential elements of addiction treatment and wilderness therapy include group dynamics and support. Within the framework of adult wilderness treatment, group dynamics are essential for promoting a sense of belonging and cooperation. Together, they overcome obstacles in the wilderness that may be similar to those they encounter in their daily lives. People can witness how their activities and interactions affect other people by receiving support and criticism from their peers in a group dynamic environment.

The dynamics of group therapy in addiction treatment give patients the chance to interact with others and get affirmation from those who have experienced similar things. Since group support provides a sense of understanding and belonging that is hard to get elsewhere, it can be especially effective. Adults can talk about their triumphs, anxieties, and experiences in these groups and get advice and support from professionals who are qualified therapists as well as from peers. An individual's chances of effectively conquering addiction can be greatly increased by this group assistance.

CHAPTER TEN

PARTICULAR POPULATIONS

VETERANS AND WILDERNESS THERAP

Wilderness therapy is becoming known as a successful way to help veterans with the special difficulties they may encounter while they adjust to civilian life. Veterans undergoing this type of treatment can overcome challenges like post-traumatic stress disorder (PTSD), depression, anxiety, and the challenges of reintegration by combining outdoor activities with psychological and emotional support. With activities like hiking, camping, and group therapy sessions, the wilderness setting offers a setting that promotes personal growth and healing.

The sense of camaraderie and shared experiences that veterans receive from wilderness therapy is an important component that can help counteract the feelings of separation and isolation that many veterans feel after leaving active duty. Veterans can connect with one another's experiences, and the outdoors provides a secure setting in which they can talk openly about their difficulties. Furthermore, the physical difficulties and successes in the environment can foster a sense of purpose and accomplishment by boosting confidence and self-esteem.

COUNSELING FOR PEOPLE AT RISK

Counseling for people at risk includes a broad spectrum of interventions intended to offer support and direction to those who are more likely to engage in harmful or self-destructive activities. These people might be in danger because of things like drug misuse, dysfunctional families, juvenile delinquency, or mental health concerns. For those who are considered to be at risk, treatment aims to pinpoint the underlying reasons for their dangerous behavior and equip them with the knowledge and resources necessary to make wise decisions.

To address both interpersonal and personal difficulties, counseling and therapy sessions for people who are considered to be at-risk frequently incorporate both individual and group treatment. Counselors and therapists try to provide a secure, accepting environment where clients can explore their ideas, feelings, and experiences. Counseling methods like dialectical behavior therapy (DBT) and cognitive-behavioral therapy (CBT) are commonly employed to assist clients in acquiring coping mechanisms and emotional control. These therapies have the potential to be very successful in stopping more harmful behaviors and promoting personal development.

THERAPY FOR PEOPLE WITH DISABILITIES

People with physical, cognitive, sensory, or developmental disabilities can benefit from therapy, which is a broad field that strives to enhance their general well-being and quality of life. Enhancing mobility, independence, and communication are the main goals, and the psychological and emotional elements of living with a disability are also addressed. Among other therapies, this one may involve speech, occupational, physical, or psychological counseling.

When it comes to helping people with disabilities regain or enhance their physical function, strength, and mobility, physical therapy is essential. The goal of occupational therapy is to assist people in acquiring the abilities required to carry out everyday duties and engage in worthwhile activities. For people who struggle with swallowing or communicating, speech therapy is essential. Psychological treatment and assistance are also necessary to manage the social and emotional difficulties that frequently accompany disability.

CHAPTER ELEVEN

SAFETY AS WELL AS RISK CONTROL

SAFETY PROCEDURES

When it comes to safeguarding the health and safety of workers, customers, and members of the public, safety protocols are crucial. These protocols are set practices and policies intended to reduce the possibility of mishaps, injuries, or other possible dangers. They offer a methodical framework for seeing possible dangers, putting preventative measures in place, and reacting appropriately when anything goes wrong. There are many different aspects of safety regulations, such as public safety, environmental safety, and occupational safety.

Using personal protection equipment, operating machinery safely, handling hazardous products, and preventing accidents are some examples of safety protocols in the context of workplace safety. To guarantee compliance and ongoing development, these protocols frequently involve written procedures, training courses, and routine safety inspections. To reduce the danger of accidents and occupational illnesses, a manufacturing plant, for example, may have safety procedures in place for the correct operation of machinery and the handling of chemicals.

QUICK REACTION

A crucial component of safety management is emergency response, which focuses on how a company handles unforeseen and possibly hazardous circumstances. Natural catastrophes like earthquakes or floods, as well as incidents at work, fires, chemical spills, and medical crises, can all fall under this category. A clear and well-executed emergency response plan will greatly lessen the possible injury and damage that could arise in such situations.

Procedures for personnel evacuation, first assistance, incident reporting, and coordination with pertinent authorities, such as fire departments or medical services, are all part of an efficient emergency response plan. All staff members must participate in regular training exercises and drills to make sure they all know what to do in case of an emergency. It is imperative to customize emergency response plans to the unique risks and requirements of a company, considering variables such as industry, geography, and the type of potential events.

EVALUATION AND MITIGATION OF RISKS

A key step in safety and risk management is risk assessment and mitigation. It entails identifying possible risks, assessing their impact and likelihood, and then creating plans to lessen or manage those risks. Organizations can make well-informed decisions on emergency response strategies and safety procedures with the use of risk assessments. Preventing mishaps, safeguarding individuals and property, and reducing monetary and image harm are the ultimate objectives.

Different strategies, such as quantitative or qualitative approaches, can be used to conduct risk assessments. Qualitative approaches rely on the expertise and judgment of experts, whereas quantitative methods use data and statistics to evaluate risks. Risk avoidance, risk reduction, risk transfer, or risk acceptance are some examples of mitigation measures when risks have been identified. For instance, a company may spend money on redundant systems to lower the chance of a catastrophic breakdown or on insurance to shift the financial risk of specific occurrences.

The essential elements of safety and risk management in any firm are emergency response plans, safety protocols, and risk assessment and mitigation. Together, they provide a thorough framework for recognizing and managing possible risks, handling crises with efficiency, and eventually guaranteeing the security and welfare of all parties involved. Adherence to these principles by an organization not only safeguards people and property but also enhances its overall viability and accomplishments.

www.ingramcontent.com/pod-product-compliance
Lightning Source LLC
Chambersburg PA
CBHW060004300526
45794CB00003B/1086